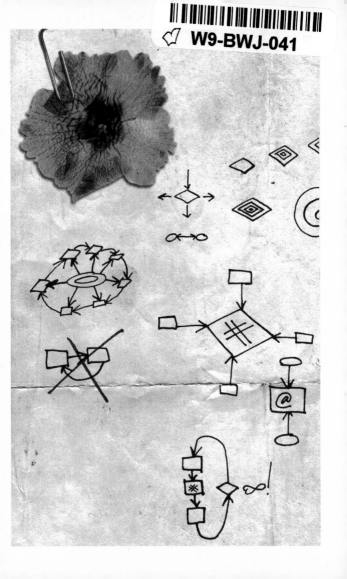

ITERATING GRACE

HEARTFELT WISDOM AND DISRUPTIVE TRUTHS FROM SILICON VALLEY'S TOP VENTURE CAPITALISTS

KOONS CROOKS

INTRODUCTORY ESSAY BY
ANONYMOUS

FARRAR, STRAUS AND GIROUX   NEW YORK

For Steve DeLong

Farrar, Straus and Giroux
18 West 18th Street, New York 10011

Library of Congress Control Number: 2015951061
Paperback ISBN: 978-0-374-53664-0
E-book ISBN: 978-0-374-71530-4

Our books may be purchased in bulk for promotional, educational,
or business use. Please contact your local bookseller or the
Macmillan Corporate and Premium Sales Department
at 1-800-221-7945, extension 5442, or by e-mail at
MacmillanSpecialMarkets@macmillan.com.

www.fsgbooks.com • www.fsgoriginals.com
www. twitter.com/fsgbooks • www.facebook.com/fsgbooks

1   3   5   7   9   10   8   6   4   2

Dear Reader:

What is this tiny book? Who is this Koons Crooks?

Privately printed copies of *Iterating Grace* began landing on the doorsteps of key figures in the technology world in the summer of 2015. Some were hand-delivered, some arrived in the mail; they were packaged discreetly and unassumingly, and arrived untraceably. Who sent them was unknown. Who wrote them was unknown. But with just 140 tiny copies, the anonymous story of Koons Crooks—a man who took the public musings and aphorisms of the Silicon Valley elite too much to heart, with unfortunate, unexpected consequences—had the tech world atwitter. It was, in the words of Alexis Madrigal, editor in chief of *Fusion* and one of the first people to receive a copy, "a perfect little skewering of the current moment."

Attention immediately turned to unmasking the creators of *Iterating Grace*. There were competing efforts to apply algorithms to the prose that promised to scientifically identify the author. Blog posts and endless comment threads pointed fingers at one bestselling, prize-winning writer after another. A consensus emerged that it must be the tip of a bold marketing scheme by a mega-corporation or an especially savvy start-up.

But gradually it became clear that it was simply this: a small piece of literary art, carefully crafted and perfectly pitched, bemused and a bit outraged—redolent, perhaps, of a twenty-first-century Mark Twain—by a creator who did not want to be identified and would not explain anything beyond what the satirical fable said for itself. Even *we* still don't know who wrote it; we have corresponded only through intermediaries and via pseudonymous email accounts. The "Friends of Koons," they call themselves. But that is not the point. The point here is *Iterating Grace*, finally allowed to speak for itself to the whole wide world.

—The Editors, FSG Originals

ITERATING GRACE

UP THE MOUNTAIN, DEEP INSIDE:
THE ECSTATIC EXPLORATIONS OF KOONS CROOKS
BY ANONYMOUS

You don't need to know my name. What's important is that I recently got a phone call from a young man outside Florence named Luca Albanese. (That's not his name, either.) Luca and I hadn't communicated in ten or eleven years; he had lived with my family in Houston as an exchange student during high school. On the phone, he told me that he'd just finished veterinary school—a poignant swerve, it seemed, from his father's expectation that he take over the family butcher shop—but it was immediately clear that he had no interest in small talk. He'd had an extraordinary experience, he said, and was in possession of "some unusual materials" that he thought I should see.

Luca and his brother had been trekking around South America and, while mountain biking on a Bolivian volcano called Uturuncu, discovered a dead body: an apparent hermit, lying outside a yurt. The scene and circumstances were peculiar—we'll get to that—but what first struck them was that the man was clearly not a local. He was white, relatively young, and completely naked, though there were some clothes folded neatly beside him: a pair of neoprene cargo pants and a tattered fleece vest that said "Pixelon" on the left breast. There wasn't much in the yurt, but Luca furtively gathered some books and papers before descending and alerting the authorities. His mediocre English kept him from deciphering what he'd found and comprehending this man's story fully, but he'd gleaned enough to decide that because I teach poetry writing and live in San Francisco, I was—at least in Luca's mind—the obvious person to put on the case.

Luca's package arrived a week later. Here is what I learned.

* * *

Koons Crooks was an inexhaustible foot soldier of the first dot-com boom, working steadily as a programmer at a string of forgotten start-ups in the late nineties: Naka, InfoSmudge, BITKIT, Popcairn. He was tall and muscular, with a square jaw, an olive complexion, and thick, wavy hair just long enough to tie back in a ponytail—and yet, several friends told me, he still always managed to look distressingly unhealthy. He was, as one acquaintance put it, "fully post-meal," inserting pieces of food into his mouth at regular intervals while he worked. A coworker remembered Crooks moving through an entire bag of frozen shrimp gyoza in a single morning, raising and lowering his left hand hypnotically and gumming each dumpling until it softened enough to be chewed and digested. Occasionally, he could be heard saying, very quietly, a single word: "Yum."

People called him Pepper, a nickname he carried from company to company until no one could remember exactly how it had originated. (One programmer told me, "I always thought it was, like, 'This guy spices things up!'") He had few real friends but was close with his Bernese mountain dog, Smoot, at whom he was known to shout commands in Unix.

When the bubble burst, Crooks left Smoot with an uncle in Danville and started driving around the West by himself, visiting national parks and Waffle Houses. It was a period of intense reflection. Gradually, he seemed to work up a metaphysical perspective on the dot-com collapse. In an email to a former coworker, he wrote:

We were like an army of ants! Think [of] the way they build and carry. We were busy building and carrying. We were building and carrying nothingness. I was never happier.

Another friend remembered Crooks calling in the middle of the night from a Waffle House near Denver, in early 2003, to share his ecstatic realization that, as he put it, the dot-com economy had been "one big mandala," referring to the Tibetan Buddhist tradition of creating art out of sand and then wiping the artwork away.

In Silicon Valley, so many people had collaborated to create so much, only to watch it crumble and wonder how real it had ever been. For a lot of tech workers, this only led to despondency and debt. But Crooks seemed to find the dot-com economy's impermanence electrifying. Start-ups, he realized, were a kind of spiritual exercise. He wanted to live that experience again, but in its purest possible form. He wanted "to touch THE ESSENCE without gloves," as he put it in the summer of 2003, in an email to his uncle.

After that, he disappeared.

* * *

Crooks's life in the Bolivian highlands remains mostly inscrutable. It's unclear when he began living on the volcano, or how often he came and went.

He seems to have resurfaced periodically back in California, though these sightings are hard to confirm. For example, while reporting from a tech party in San Francisco in February 2014, the *Re/code* reporter Nellie Bowles noticed a man kneeling quietly next to a mechanical bull, stroking the fake animal's flank with his eyes closed. When Bowles introduced herself, the man gave his name as "Koons" but declined to be interviewed for her story. Bowles—whom I reached out to—could not say for sure whether the man communing with the mechanical bull fit Crooks's physical description.

If you try
to outlaw
the future
it will just
happen
Somewhere
Else.

— Paul Graham

Meanwhile, the venture investor Chris Sacca recalls a man named Koons approaching him on a ski slope in Truckee in either 2011 or 2012, asking for a meeting. The man had no skis on, or even boots. When Sacca asked for his contact information as a way to politely wiggle out of the confrontation, the man pulled a tattered business card out of his wallet and used a Sharpie to cross out the name and details of the Palo Alto-based podiatrist on it, then paper-clipped a small dried wildflower to the blank side, put the card in Sacca's hand, and wandered off into the snow.

Beyond food and camping gear, Crooks had very few personal items with him in the yurt: one change of clothes, a solar-powered laptop and satellite Wi-Fi hotspot that he seems to have built himself, an iPod shuffle with exactly one song ("Even Flow"), a photograph of Vannevar Bush, and a small, utterly inexplicable selection of hardback books: four volumes about deep-sea exploration (including one by a former roommate, James Nestor) and a copy of *Chez Panisse Vegetables*.

These books are what Luca recovered and sent me. They are astonishing artifacts: Crooks annotated their pages with unrelenting ferocity, filling virtually all the white space with emphatic philosophical treatises or disjointed bursts of observation; it was as though he were reading, and responding to, other books entirely. In fact, he had crossed out the titles on their covers and scrawled the same new one on each: "Iterating Grace."

Other diary-like entries in the margins describe Crooks's pared-down, monastic lifestyle. He took walks. He foraged. He dried wildflowers and grasses. He collected interesting rocks and numbered each one sequentially with a Sharpie. He devoted much of his time to contemplating ecology, which he called "the ultimate operating system." And with increasing frequency, it seems, he checked Twitter.

Crooks's family recently deleted his account, but he seems to have followed a long list of founders, venture capitalists, and angel investors. He began to see, in their Tweets, hints of some elusive but irrefutable wisdom: a string of logic that underwrote the universe like code. For him, the tossed-off musings and business maxims of these men (they were almost all men) shimmered with a certain numinous luster. He contemplated individual Tweets for days, sometimes weeks, expounding on them at length in the margins of his books about the sea, meditating on them as though they

Experience + Curiosity
= Powerful.

Experience − Curiosity
= liability

The world is dynamic.

− Bill Gurley.

Good
Stories

always beat

Good

Spreadsheets

— Chris Sacca

were koans. The answers he'd been searching for had been there, in the Bay Area's innovation economy, all along—articulated, unwittingly, by an elite class of entrepreneurial high priests. Crooks could see that now from his new vantage point, 5,300 miles away from Silicon Valley and 12,000 feet in the sky. "Angel investor = herald, messenger. Message = Hello, World!" he scrawled beneath an illustration of beets in *Chez Panisse Vegetables*.

Slowly, he began to write out the Tweets himself in various half-formed styles of improvised calligraphy. (He apparently became interested in the art form after reading that Steve Jobs was a practitioner.) For months, Koons Crooks transcribed Tweets, alone on his volcano. He worked on large sheets of paper, then pinned the papers around the yurt, where Luca eventually discovered them. They are here with me now, as I write this: all these aphorisms, spread on the floor of my office—a patch of fertile ground from which one man's spirit slowly sprouted and flowered.

Crooks himself never tweeted. He never retweeted. But he had developed his own impassioned way of favoriting.

* * *

Maybe you've heard the gossip around the Valley and think you know how this story ends. But, as I discovered, there are a lot of misconceptions about the llamas.

First of all, they weren't technically llamas, but vicuñas: gangly, small-headed camelids of the genus *Vicugna*. Vicuñas are reclusive by nature, but scattershot entries in the margins of Crooks's books describe the animals becoming increasingly acclimated to him. I've since read that it's not unheard-of for wild animals, such as mountain goats, to approach hikers and lick them; they're attracted to the salt of their sweat, like a salt lick. Crooks's relationship with the vicuñas started with a similar phenomenon, except that the animals were drawn in by the salt of his tears.

They were tears of joy, it seems: the longer Crooks was isolated, the more deeply, and eccentrically, his contemplation of the Tweets became, and the closer he felt he was coming to a variety of enlightenment. He describes sitting outside his yurt for hours at a time, weeping—his soul swirling inside some kind of uncontrollable ecstasy. And when the vicuñas began to lick him, he writes, his weeping turned to laughter.

You can fool some of
the people all of the
time, and you can
fool all of the people
some of the time, but
you can't fool all of
the people a

- Rick Burnes

Reminder to Self:

not happy with game?

change the game.

— Brad Feld

You may have heard that he was trampled by llamas in a freakish accident. This, too, is inaccurate—and I say this based on Crooks's writings before the event and also on photographs of the scene that Luca sent me and which, owing to their brutality and rawness, he now regrets taking. (Luca feels strongly that I mention he's since pleaded with me to destroy the photos; I have refused.)

Yes, there were many deep, hoof-sized contusions on the head, face, and chest; broken ribs and a fractured skull. And yes, there were vast swarms of bites and abrasions—presumably from the animals' wiry-haired muzzles. But there was no sign of any kind of struggle or resistance: in fact, Crooks's body was positioned on its back, with its arms and legs out at perfect angles, like Leonardo da Vinci's drawing of Vitruvian Man; he seemed to have simply undressed and lain down. Then, beyond that, the many open, empty cans; the traces of the cat food he'd apparently smeared all over himself.

People who knew Crooks are, of course, devastated. He seems to have acted from a place of supreme happiness and clarity, and yet his actions have only caused sorrow and confusion for the people he left behind.

Happiness? Clarity? Are we sure this was his mind-set? I think we can be. Because there is one other detail: a small, badge-like piece of paper, which Crooks had crudely laminated with plastic wrap and tied around his neck with a lanyard, as though he'd just registered at his final conference. By the time Luca found him, vultures or other scavengers had taken one of Crooks's eyes and most of an ear. Elsewhere, maggots were doing their work.

Luca reports that Crooks's body had already started to decompose to such a degree that, in certain places, the boundary between it and the soil beneath it had blurred. Koons Crooks—Pepper—was becoming granular: converting into dust and earth and nutrients, just as he apparently intended. His broken mouth was still smiling, and on that little sign around his neck, he'd written:

THE SHARING ECONOMY ;)

bad news :
no WIFI
on this flight

good news :
no WIFI
on this flight

– Seth Levine

More Startups Die
of Indigestion
than Starvation.

Focus Wins.
Doing too many things
fails.

— Bill Gurley

old:

all software expands
until it includes
messaging

new:

all messaging expands
until it includes
software

~Benedict Evans

People who spend money
on things I don't value
are just showing off.

People who don't pay
for things I value
have no taste.

— Benedict Evans

EDITOR'S NOTE: At first, I was confused by the conspicuous shallowness of this Tweet. Its perspective seems limited, even petty, while the others are expansive. Was Crooks—a man without a shred of cynicism—mocking Evans? I can only assume he read this as wryly self-aware: Evans laughing at both his own judgmental impulses and the narcissistic limitations of all human consciousness.

Happy Passover to all!
Message still resonates:
count your blessings,
appreciate what you've got,
stop wandering &
go build something great.

—Jeff Bussgang

It's crazy how each
iteration of product is a
ton of work,
but makes @grasswire
more simple,
not more complex

— Austen Allred

Reminded again
this morning,
that there's
more to life
than Rushing.
Blue Bottle
offers
us all
a precious reminder
to Slow Down...

~ Tony Conrad

the decentralization
of
everything

tmblr.co/Z9mGay1E47F28

– Andrew Parker